WHY NOT BELIEVE?

WHY NOT BELIEVE?

And why I believe . . .

ROGER MALSTEAD

with RONALD CLEMENTS

Columbus, Ohio

Why not believe? And why I believe . . .

Published by Gatekeeper Press
2167 Stringtown Rd, Suite 109
Columbus, OH 43123-2989
www.GatekeeperPress.com

Contact Address:
Mr Roger Malstead
rogermalstead@gmail.com

Dr Ronald Clements
Mount Cottage
42 Telston Lane
Otford
Kent
TN14 5JX
01959 522730
clementsntproject@hotmail.co.uk

ISBN (paperback): 9781642372656

Printed in the United States of America

Cover arranged by John Barrett

About the Authors

Roger Malstead was the founder and director of Spear Publications, an organization committed to translating Christian resources for churches in Turkey. He is also the producer of feature films (*Bedel, Ali and Silvana*), and more recently a series of TV documentaries (*The Jesus Accounts* [2010], *Jesus: Son of God?* [2014], *Jesus: Dead and Buried?* [2018]). He and his wife, Yvonne, have lived in Beirut, Turkey and the UK. Now retired, they live in the USA.

Dr Ronald Clements is a full-time freelance writer living in the UK. His biographies include *Lives from a Black Tin Box* and *In Japan the Crickets Cry*. He has collaborated with Roger Malstead as director and scriptwriter of the TV documentaries *The Jesus Accounts, Jesus: Son of God?*, and *Jesus: Dead and Buried?*. www.ronaldclements.com

Contents

About the Authors.. v

Where in the world should we begin? ix

1. Why not believe? 1
2. Starting with Something.. 3
3. A Beginning of Hope 7
4. God Options.. 11

Before I go on 17

5. Moving On 19

Back in the day . . . 1.. 25

6. A Piece of History.. 29
7. Jesus: Spot the Difference?...................................... 33
8. Death and Hope.. 37

Back in the day . . . 2.. 43

9. Getting up on Sunday 47
10. This Book: The most valuable thing that
    this world affords 53
11. On Word 57
12. God's Word.. 63
13. God Calling 67

Finishing with Something 71

Things to read 73

Where in the world should we begin?

WHERE IN THE world should we begin? *I'm just asking that question because the world is a big place. And there are a few billion people in it. Chances are, like you, I've met several thousand of them one way or another. Family, friends, work colleagues, the folk at my church, I know very well. Most of the rest have just been 'hi and goodbyes'. In between there have been hundreds of acquaintances, whom I enjoyed knowing for a while. And these days I have dozens of contacts whom I never see, but who send emails, texts and use whatever social media is in fashion to tell (and sell!) me stuff.*

The Internet reckons I might meet 70 to 100,000 people in my lifetime. And since I am in my seventies I must be getting close to my quota. Who knows? What I do know is that people, like you, are important, no matter who they are and where I meet them. Which is why I wanted to write this. For family, friends, acquaintances and… if you are one of the many people I've not yet met, well, for you too.

But first let me introduce myself.

I was born and grew up in the United States—Nebraska; just north of the middle if you are looking at a map. My family lived in a few places, including California. Then I went east to Illinois for college and back to California for a while. After that things changed big time. I headed for Europe.

That was when a mate of mine and I got arrested in the Soviet Union. And when I got escorted out of Turkey and dumped at a border post in Greece. The stories are here if you're interested. But why did these things happen? Well, I was keen to tell people what I believe and sometimes you meet guys who… OK, they just don't want to boil up zavarka, or get a Turkish chai, and give you the time of day.

But some do and you learn a lot about what they believe. People in Turkey and the Middle East, where I have also lived, wonder about a God who is a long way away, but who has a lot to say about how they live their lives. In the West generally folk doubt whether a God exists at all and are trying to work out where they fit in the world without one.

Me? What do you believe, Roger? And why?

I have spent my adult life answering that question—'what do you believe?'—for all kinds of folk, across the States, down through Europe and on into Asia. Actually, it's an easy question to answer. I know exactly what I believe and, as you gather, I am more than happy to tell you.

But the why? Well, that takes a bit longer and most people don't have time to listen to old guys. So I decided to write it down, so you and others can get a coffee, pick it up when you want to, and maybe think about it when I am not around.

1

Why not believe?

THERE IS A verse in the Bible that goes like this:

Always be prepared to give an answer to everyone who asks you to give the reason for the hope that you have.[1]

It is quite reasonable. Everyone believes something about life and why we are here in our world. A lot of people have strong views, others just drift into some kind of belief. Mainly everyone thinks they have got it right or right enough and as long as no-one challenges them about it, life goes on.

In the Middle East I found plenty of people who were worried that if they questioned what they believed, their world was going to collapse in a heap around them. In the West it's much the same; folk actually aren't that different. We prefer to read articles, books and websites, listen to people that see things the same way as ourselves. Social media repeatedly pushes us stuff it knows we 'like'. But sometimes . . . just now and again . . .

---

1   1 Peter 3:15 (NIV).

I wonder why we believe what we do. And it is good to go back to the basics, see what got us to where we are now, and where we might want to go next.

When I say basics, that's exactly what I mean. But let me start with this. When it comes to what people believe, we can be big on logical argument, suspicious of experience. This, of course, is a weakness, not a strength. We have minds and emotions. We have to ally both for a truly fulfilled life. This is why I have included some stuff about my life along the way, so you can see that what I believe isn't just something confined to a book. That said, some people would say that my faith cannot stand up to the scrutiny of logical argument. I guess they are just plain wrong . . .

2

Starting with Something

THE NEATEST THING about living is that we exist. I exist. You wouldn't be reading this if you didn't exist. I have a house with a front lawn and a bird box that birds visit. Across the street there is a lot more grass and some trees. When I open the curtains in a morning, I can see the mountains. I drive down into town and find stuff everywhere I go. I open my computer or look at my phone and, hey, people I can't see communicate with me and send me pix of their homes and their stuff.

Sure, there are some people who say we don't exist at all. It's a dream or, if we are more 21st century, we live in a giant computer programme that churns away in a vast experiment orchestrated by who knows whom. Science would tell us otherwise. And I am going with science here. Plus, when I stub my toe or decide to take a shower, the reality of my existence looks in good shape to me.

So let's start with *Something*. And let's go big. The universe is the biggest something we know and it contains all the somethings we can observe, in whatever way we can do this. It also contains things we know are there, even if we can't currently measure them, and we guess that there are things there we have yet to discover. But all in all, our universe exists.

[Text Box]

> *An aside: The multiverse—that's when several universes, maybe millions and more, each with its own characteristics, all exist together (ours being just one of the many)—is an interesting idea, but it is a theoretical entity. Maybe I will come back to that.*

First up: A big universe and a big question: Why is there something, not nothing?

There are only two ways that the universe can exist:

- because it was made.
- because it has to.

The first is easy to understand. The meals on my table each day are there because Yvonne (a far better cook than I am) went into our galley kitchen, found ingredients, cooked them and called me to the dining room to eat. Without Yvonne they wouldn't be there. Same with the universe. Someone/something caused it to be made, so it exists.

The Big Bang is a scientific explanation for how the universe started. But, whether you believe in 'Mother Nature' cooking it up or some guiding design in it all, there is still the understanding that somewhere back down the line someone/something caused it to exist in the first place.

Unless, of course, you think the universe exists because it 'has to'—it *exists necessarily*—it is in its own nature to exist. So you can perhaps think of mathematics. Numbers just are. Nobody made them. So is the universe like that?

The truth is few people, if any, believe this. Atheist or theist. Any science programme you watch or any lecture you go to on astro/quantum physics will be based on the basic belief that stuff is brought into existence by other stuff causing it.

Now here is the point: what/who kicked it all off in the first place? The simple answer that theists give is 'God' (leaving aside who that God may be and how he/she/it did so for the time being). And atheists tend to rush to a standard response, which pretty much is that science has put the lid on that notion. God isn't needed.

Now *not* believing in God is fine, until you are asked to give a reason for it! And when it comes to the existence of the universe, the neat answer that science has sorted it doesn't go back nearly far enough. The best that can be said is that the universe exists without a reason. Heaven help us! Imagine how a believer would be ridiculed if they couldn't find a reason for their faith.

And this is where teenage philosophy kicks in. 'OK. If God exists, who made God?'

Like the universe there are two ways God can exist:

- because God was made.
- because God has to.

God, unlike the universe, can exist without being made by someone/something else. God can *exist necessarily*. Is that unreasonable? Well, it depends what you think God is like. Lots of ideas of God may not work when it comes to the existence of the universe. But then, some do.

> *An aside: Saying God 'has to' exist may sound like God is somehow being forced to exist. That's not what I mean. This is about the essential nature of God.*

Try this: if God wasn't made by someone/something else and it was God that kicked off the universe, God cannot be the same as the stuff of the universe. Someway God must be different. So, in what ways does he/she/it need to be different?

The basics of our universe are matter, space and time. So, let's be as different as we can be. God must be non-material. God must not exist in physical space. God must not be governed by time. This kind of necessary God can cause the starting of something—like the universe. Now, isn't that neat?

3

A Beginning of Hope

MAYBE THE QUESTION you are asking is—OK, so God can exist without being made and maybe the universe can't, but what if the universe didn't have a beginning and won't have an end. It has just existed in one shape or another forever; a long, long string of causes producing something that has resulted in our universe?

At this point we go back to science again. Would you please welcome the Big Bang!

An aside: Credit for proposing the Big Bang theory goes to a Russian scientist, Alexander Friedmann, and to a Belgian Catholic Priest, Reverend Monsieur Georges Lemaître, working independently. Friedmann and Lemaître made their discoveries in the 1920s, 90 years ago. The basic theory has survived a barrage of advances in astronomy and physics, and it looks like it is going to be around for a good while longer.

The Big Bang is a great name for a scientific theory. The popular press love it and the public think they understand it. After all, if you are going to start off a universe, a big bang should do it. Unfortunately, the Big Bang was not big and there was no bang. It was not an explosion. It was a rapid expansion. An expansion around 13.8 billion years ago of all the matter that exists from (depending on which Internet article you read) something the size of a peach, or pinhead, or a 'singularity'—a 'space', where the distance between any two points is zero. If it helps, imagine every last particle in our universe crammed down into the smallest 'speck' you can think of. A space so densely packed that sound and light waves are going nowhere, because there is nowhere to go.

Another common misconception is that there was a huge space outside this tiny 'speck' into which all of everything expanded. But this isn't right either. The expansion was within itself, not into something. That is harder to imagine, of course, because we have balloons that expand into the air around them and we understand them. However, if you were a creature living in only two dimensions within the surface of the balloon, you'd have no problem understanding that your 'universe' was getting bigger, without worrying about what it was getting bigger into.

And, just in case you wondered, the Big Bang also confirms that this beginning of space is also when time began.

Enough textbook physics!

The point here is the Big Bang tells us that the universe had a beginning, when this singularity enlarged rapidly and particles started to shape the expanding cosmos we see today. Space, matter and time, the ingredients of our universe, all started then and there.

Why is this important? Among other things it raises the

question for science: why did the Big Bang happen at all? Any self-respecting scientist will put up her or his hand and say we don't know. Because that singularity 'speck' was so dense, we have not been able to observe anything in our universe that will give us a clue. Before light was able to travel into the spaces between the matter, we can only guess what went on.

Some people have claimed at this point that the universe performed a really neat trick. It started itself. The very scientific equivalent of you grabbing your shoelaces and lifting yourself off the ground. Good luck with that!

So, let's go back to God. The universe had a definite beginning. Therefore something/someone had to start it off. Even if we cannot work it out, there has to be a cause, otherwise the universe would not exist. Believing that a God, who is altogether different from the universe, could be the cause is not unreasonable, whether you agree or not. Otherwise you have to simply say we can't know and leave it there for eternity. Or believe in bootstrap magic.

Another important thing here is that having a beginning gives us hope. In contrast, a universe that has always existed, in which matter is continually shaped and reshaped, pieced together by nothing more than the physics and chemistry of its own inherent nature, ultimately provides no purpose to the lives we try to lead here and now.

Similarly, science paints a bleak future for the universe—a 'heat death'. Given enough billions of years, bloated, expanding space will become a thin soup of exhausted energy, a stagnant sea of joyless equilibrium. Not a lot of hope to be had there!

If there is a God who transcends all of this, I have a hope. I can understand that there may be some purpose in living. However, for that, this God needs to be something more than non-material, unbound by time, existing outside the confines of my universe. He/she/it needs to have intelligence; an ability to give shape and purpose to the universe and to my life.

OK, you guessed, that's the line to get you to read the next chapter . . .

4

God Options

OUR WORLD IS a beautiful place.

When was the last time you caught the silver of frost on a spider's web in the early morning light? Or studied the patterns of ice glazing a hollow on the path ahead of you? Perhaps you prefer the strength of the sea crashing into rocks, catapulting waves high into the air. Or catching sight of an eagle rising on unseen thermal currents. Have you walked deep into a rain forest? Or stood on the very tip of a mountain? Have you stopped and looked and listened and encountered the beauty around you? The earth is a good place to live.

Splendour, of course, doesn't end with the earth. No two sunsets are ever the same. And we are privileged now to be amazed by stunning photos of distant stars and huge galaxies, icy comets and dusty nebulae. The universe, we have discovered, is also a beautiful place. And remarkably it all works just fine.

Generally, we take the fact that things work for granted. Until they stop working. If the universe stopped working, we would definitely know about it. But it doesn't. Unaided by us it continues to exist, continues to function. Immensely complex. Incredibly elaborate.

What is more extraordinary, according to eminent scientists, is that all this works within a small margin of error. One of these distinguished experts has identified six numbers which help keep the whole thing functioning. One is the strength of the force that holds atoms together related to gravity. Another determines how well the nuclei of atoms bind together. The third measures how much material is stacked up in the universe. The fourth controls the expansion of the universe and the fifth its 'texture', whereas the sixth is something we all know. We live in three physical dimensions, not two or four. (Time is a different kind of dimension.)

> *An aside: These 6 numbers come from the Astronomer Royal in the UK, Martin Rees, a very prominent scientist. Others writing on the same subject may identify different factors, but the conclusion is the same; our universe operates on finely tuned parameters.*

Make a tiny variation in any of these numbers and there goes 'our' universe. A universe, if it did actually exist, could still scrape by, but 'not as we know it'. Furthermore scientists show that 'life' in any form in this alternative universe—the ability to feed, adapt and reproduce—would be impossible. If you were wondering if a different type of life-form would evolve and live happily ever after, forget it. It's not an option. Not a hope.

So where does this leave us? We have a very finely-tuned universe that works, started by something 13.8 billion years ago. How did we get from there to here? Well, there are choices on what to believe.

If you don't believe in God, then God obviously isn't an option. But, there still has to be some way the universe keeps ticking away in its own space and time and doesn't disintegrate into a lifeless whatever. A 'no-God' explanation is required; a 'natural', rather than a 'supernatural', explanation, if you like. Then, the argument goes, if we have a 'natural' explanation and this doesn't require God to be involved, God doesn't exist. Unfortunately, this has been presented as the 'scientific' and therefore the 'right' solution. End of story.

However, don't those who hold this view need to explain why they believe this? Was it simply that they didn't believe in God in the first place and adopted this 'scientific' option as the most convenient one on offer? Is calling it 'scientific' a way of avoiding the most obvious question for a scientist—where's the evidence? At best, the no-God option remains an option—a belief system in its own right, to be challenged, just as people ask me to give a reason for my beliefs.

It is worthwhile examining the no-God option, because it is not as strong as its believers would tell us. The standard answers available are that the universe has developed into the life-permitting universe we enjoy because:

- it is a physical necessity or . . .

- chance threw up the right parameters.

The trouble with the first is that universes that don't allow life are very many and those that do are very few—hence the fine tuning. So somewhere in the Big Bang initial state there must have been a preference (a necessity) for our universe. But why? Nothing says it must have been that way. In fact, the opposite is true. It makes more sense for there to be a reason for a preference, than for it simply to exist as a preference.

Which takes us to chance. Some universe had to be produced and, hey, we were lucky enough to get this one!

Chance sounds a good explanation until we ask why we were lucky. If the odds on winning are truly very small, then there is always a nagging doubt. Did we win because someone rigged the whole thing? How can we be sure? The truth is we can't.

The fine tuning of our universe also puts us in a pickle. Everything tells us there was a very slim chance that our universe exists at all and a very slim chance that it should work. The odds are not stacked in favour of a no-God option. If anything, believe it or not, it is more reasonable to consider that not only did God kick things off, he purposely made it to work this way.

OK, OK. I need to talk about the multiverse; the idea that there is not just one universe but millions upon millions—or even an infinite number. If there is enough of these, surely one—ours—can exist without worrying about God. In effect, every kind of universe wins.

This again runs into problems. First and foremost, who knows other universes exist? We have no way of knowing. So we can say we believe they exist, but it proves nothing. Secondly, if millions of universes exist then some valid, possibly finely tuned, mechanism to generate universes must also exist. And so we go back to the same questions we began with—why should this mechanism exist?; the mechanism must have a starting point—how did it start?

The only other option available is to assume there is some amazing power which underlines the existence of our wonderfully working universe. Since we understand that intelligent minds exist in our universe—you and I both have one—the most logical conclusion is that this power has one too. A smart 'Grand Designer', if you like. This 'Designer' brought all things into existence and keeps them existing.

Inevitably, there are differing views of a 'Designer'. Did he/she (if God is intelligent, I think we can drop the 'it'!) put the Big Bang in motion and just leave it to get on by itself—an absent landlord? Did he/she intervene along the way, nudging it in certain directions—a maintenance 'man'?

Let me tell you what I believe about God. Yes, he/she is different from the stuff of the universe—non-physical, non-material; spirit. Yes, he/she exists outside of time—eternal. Yes, he/she is intelligent—omniscient, knowing everything, in fact. But I don't believe God is an absent landlord. So, is he/she a maintenance 'man'? Well, no, I believe in a God who is far more than that. He/she is more akin to a parent, who nurtures a child. God cares. God loves. And therefore he/she cradles this universe, our world, each one of us; intimately involved from start to finish.

Before I go on . . .

To keep writing God as 'he/she' is not a great way forward. God is often described as 'he', but the Bible makes it clear that God cannot be defined as being male. God is spirit and therefore neither 'he' nor 'she' is appropriate. Nevertheless, from now on I am going to go with 'he', rather than 'she', out of convention. To be clear, I am not limiting God in some way, nor insisting on a patriarchal viewpoint.

5

Moving On

Science can only take us so far. By its very nature it is limited. Some people want us to believe that science has the solution to everything. But this is a superiority complex on steroids. At one level it demeans a whole load of other valid ways of understanding life—there are many ways of knowing. At another it is just not going to help us move on from where we have got to.

Perhaps it will be beneficial to consider getting from A to B, where B is a long way from A. The obvious choice of transport would be an airplane. Let's call this aircraft 'science', because science has taken us a long way forward in understanding things. But B isn't an airport, so we need another form of transport to help us on the next stage of our journey. Maybe we need a train because B is a railway station, or bus (bus stop), or car (most other places), or maybe we need to walk part of the way (some place else).

This is not to say that science hasn't been helpful or we can ignore it. On the contrary, science is a great vehicle to ride on, giving us views we would never see otherwise. Science is fantastic at telling us about the properties of things once they turn up.

Nevertheless, there inevitably comes a point at which insisting on science as the only way forward leaves us stuck on the pavement outside the airport. It doesn't tell us how things existed in the first place, nor why. Science needs the universe to be intelligible, but it can't say why there is such a universe. We need to hitch a different kind of lift to discover what else we can learn.

Apart from science, there are three obvious places to look for help. First up—theology—the study of God. It is maybe less attractive than science for most people, but that doesn't make it unhelpful. Second, back over our shoulders to history—is that surprising? Actually, when we come to history, there are a few surprises! And third, inside ourselves—personal experience—which I have said has to be taken into account, otherwise we are liable to miss out on ourselves as part of the answer. In this booklet I have told a little of my own story, which may, I hope, interest you.

OK, this all begs another question. How can we learn anything about God? In the last chapter I mentioned that God could simply be an absent landlord, never turning up to repair the towel rail or bothering to check what we have done to the place. Logically, we can only know about God if he puts in an appearance.

This, I believe, is exactly what God has done; put in an appearance. Theology, history and personal experience are all ways to show God is involved in our world. God is not absent.

I want at some point to have a real look at the Bible and why we should bother with it, but not for the moment. There is, however, a really neat phrase in its opening words:

And God said . . . [2]

---

2   Genesis 1:3 (NIV).

These words come in one of the most famous passages in world literature—the story of the creation of the universe in Genesis chapter 1.

An aside: This Genesis account is not a scientific description as we would understand it. Not surprisingly, when this was written down millennia ago, the writer was not familiar with modern astrophysics. It is, however, an intriguing read!

The main thing here is what it says about God. God communicates. This understanding—God speaking in some way—is repeated 15 times in the first chapter. It is as though the writer wants to make sure the readers get the point. God is not silent. The writer then goes on with his narrative, repeatedly reporting God talking and doing things to interact with our world. My personal experience is that God is still speaking today, if we can be bothered to listen.

There is another characteristic of God that I want to finish this section with. The goodness of God. This is vitally important to us. If God exists, what kind of God is he—good, bad, a mix?

I hope we all agree that moral values do exist in the world, that we do know the difference between good and bad. Otherwise, we are left with a deadly void in which any manner of behaviour is valid—vile, as well as virtuous. Sure, different cultures and different ages may quibble about how these values should be lived out. But a society without positive morality—goodness—is no place to be.

Agreed? Yes?

In western societies in particular a lot of people would say that social rules are made up by people and then enforced for the sake of getting along together. Society decides upon relative values called 'good', 'evil' or somewhere in between, which suit itself and change from time to time.

It is difficult to avoid the underlying problem of relative values—a lack of objective or absolute norms—if the source of goodness does not exist outside of ourselves. Even the most benevolent of societies will struggle to define objective good and evil, if it generates its understanding from within itself. Why should its particular viewpoints be accepted as valid?

It may be argued that moral values just exist, without God. But why and how? Where does a sense of goodness actually come from? If we have a universe solely generated by particles interacting in one long complicated series of happenings, what gave those particles some moral status to bring morality into being? Scientifically that concept is laughable!

What would be more helpful is having absolute values, which don't just change at the whim of groups of people—whether that be a democratic decision of lots of them or the desire of a few guys who run things.

God comes to our rescue. Not to invent morals for us to grapple with. But because he communicates his own absolute nature to us. The Bible talks about people being:

> . . . in his [God's] own image.[3]

Therefore, if God is good, then people have an understanding of good (irrespective of whether they believe in God).

This, obviously, is a much longer discussion than I have allowed here! Let me cut to the main point of all this. What kind of God do I believe in? A good God, who communicates

---

3   Genesis 1:27 (NIV).

with the world. Now the best bit: history, I believe, shows this supremely in the life, death and resurrection of Jesus.

But before we get to that, this is a good time to talk about a bit of my own history . . .

Back in the day . . . 1

*I*WAS BORN IN *June 1941. Europe was already at war with the rise of Nazi Germany. Six months down the road Japan would fling itself into conflict with my homeland—USA—at Pearl Harbour. It would be more than three years before peace broke out and life could really begin to be normal.*

Initially, my parents and I lived near West Point, Nebraska, north-central USA, on the old farmstead where I was born. The land was all yellow-brown loess hills and wide irrigated plains; a fertile area, which exported maize, hay, sorghum and soybeans to other states. Later we moved to a town called Wahoo and then into one of Nebraska's small cities, Fremont, before finally settling in California.

My Dad was a carpenter, working in a cabinet making shop. My Mom stayed at home and cared for my younger brother and me and, much later on in California, for all three of us boys. Neither of my parents were Christians until one day a lady selling Avon products—skin care ointments and the like—knocked on the front door. Whether my Mom actually bought any of her stock, I don't know. Somewhere in the doorstep conversation God or Jesus or church got a mention. My Mom was interested enough to start going along to one of the local churches.

Once my Mom decided to follow Jesus, she began to get involved with the church's activities and my brother and I went along too. Then she started a 'Good News Club' in our home, where, as far as I was concerned, the activities, Kool-Aid (a juice drink) and cookies were definitely 'good news'. My Mom, however, had a job for me. I had to run around the neighbourhood inviting all my friends to hear about Jesus—something I have been doing ever since. Once everyone was in our house, my Mom told Bible stories using some felt figures stuck on a piece of flannel—basic, but at the time I found it all quite captivating.

Sad to say my Mom's decision did not go down well with my Dad. Suddenly, there was a lot of tension in the home. My Dad was not happy with her and us trotting off to church regularly and she was frustrated that he was not prepared to listen to what she was discovering about God and Jesus.

This went on for a while until the church put on big evening meetings to which folk were invited to find out more about Christianity. My Dad did not want to go. Somehow my mother persuaded him he needed to and somehow on the last Friday evening he agreed to go along. But it was not without protest. He refused to change out of his work clothes. So he turned up looking like he had just walked out of his workshop, covered in sawdust—which he had. When my Dad came home, things had changed. He had decided that he too needed to follow Jesus. And so, harmony in the household was restored.

One week in the 'Good News Club' my Mom showed us the 'wordless book'. No words, just colours. But telling a story nonetheless. White represented the purity of Jesus' life. Black was all about the wrong things I had done. Red was obvious—this was the story of the death of Jesus on the cross, allowing my black 'page' to become white. Gold was the promise of 'paradise'— heaven, eternal life. And green was a reminder to 'grow' into a deeper relationship with God. All very simple, but it had a *profound effect on me. I got it. And I wanted to follow Jesus for the rest of my life.*

Of course, I had to remake this decision as an adult. But this I also did as I moved into my teens and then went off to Wheaton College, a Christian college in Illinois. In those days I was into American football big time and I had offers from universities with major teams. But Wheaton, although it had only a small team, suited me better in terms of where I wanted to go with my life. It proved a good decision. It was where I met Yvonne, my wife, and where I encountered Christians with colossal visions to change the world.

At Wheaton I started going along to 6 a.m. prayer meetings. These were not just eyes down, mumble along for a while meetings. On the table were maps of the world. These were our focus for prayer. Week after week we prayed over countries, whatever their political and religious colour, wherever in the world they were.

Jesus' last words to his followers were that they would be witnesses to his life, death and resurrection locally in Jerusalem, to the Jewish nation and 'to the ends of the earth'.[4] *This did happen in their lifetimes—as far as they understood where the ends of the earth were. By the twentieth century, however, Jesus' promise had literally gone the distance.*

So, I guess, this is a good time to talk about Jesus, before I tell you the part where I ended up looking at the wrong end of a Kalashnikov for going to the ends of the earth.

---

4   Acts 1:8 (NIV).

6

A Piece of History

A FEW YEARS AGO I wouldn't have bothered to start this chapter with the question I am going to ask. The rise of the Internet and social media, where every conceivable hypothesis can be presented as fact, has made it difficult to avoid. Did Jesus exist?

I find it frankly amazing to discover that there are some very intelligent people who say Jesus didn't exist. Jaw-droppingly amazing, in fact. Sure there are lots of hangers-on to the viewpoint, but let's credit people with having thought about the topic. How on earth did they come to that conclusion?

Chickens and eggs come to mind. What motivated this egg hunt in the first place? Was it a desire to determine historical fact? Or a creative way of bolstering an existing belief system? If you can declare Jesus didn't exist, then Christianity clearly doesn't need be given any credibility. Generally, I am inclined to believe the latter reason, simply because it is difficult to dismiss the substantial historical evidence that Jesus existed without both killing the chicken and scrambling the eggs!

An aside: One of the ways people do this is to dismiss all the evidence and then announce there is no evidence that Jesus existed. Somewhere in there you will find a leap of faith—something not reserved for religious believers!

There are numerous pieces of ancient evidence that Jesus existed that I can quote:

- Detailed stories about his life, set in a real historical time frame and in known geographical locations.

- Extensive references to him and his life written within the lifetime of eyewitnesses.

- References to him as a real person by non-Christians.

- The existence of Christianity itself.

In short, you have to have some very disputable arguments to reject the historical evidence for Jesus. Honestly, it is far easier to accept Jesus did exist than to go through machinations to 'prove' otherwise.

So what does this historical evidence tell us? That Jesus was born around 2000 years ago in Israel. That he was a rabbi—a Jewish teacher—travelling mainly in the north of the country, periodically going south to Jerusalem, over a period of three years. That he was immensely popular with the ordinary folk, but that he fell foul of the local ruling elite. That about the age of 33 he was arrested by these rulers and handed over to the Romans (who had occupied the territory) to be executed. He died nailed to a cross outside the walls of Jerusalem and was buried in a tomb close by.

Of course, if this was the end of the story, it is unlikely that

Jesus would have got a footnote in history, never mind the libraries of books (including this one) and the reams of Internet pages that exist today. It is the rest of the story that makes the difference.

Consequently, it illustrates another point. If Jesus had been just an itinerant preacher with simply a talent for upsetting the status quo, nobody would go to the extraordinary lengths people do to try and show he did or didn't exist. If some of the arguments used to illustrate Jesus didn't exist were applied to events in the lives of historical figures we happily accept as genuine, we would be left with a lot of gaps in history.

So what is it that makes the difference? What is at stake that puts people in such a tizz to prove Jesus wasn't the real deal?

7

Jesus: Spot the Difference?

THE BEST WAY to find out about Jesus is to pick up a Bible, turn to the second part—the New Testament—and read the four accounts of the life of Jesus, *Matthew, Mark, Luke* and *John*. The first three follow a similar pattern in their angle on the story. *John* comes to it with a different approach.

The key things I want to point out here are the teachings of Jesus, his miracles, and what happened after his death. These are what make the differences between Jesus being a forgotten footnote in history and him being the catalyst for a globe-sized encyclopaedia of consequences.

You can read the stories for yourself. All I would ask is that you read them with an open mind. One of Jesus' great lines to his followers was:

Come and you will see . . . [5]

---

5   John 1:39 (NIV).

It was a simple invitation to them, no pressure. If you are interested, Jesus says, come along and see who I am. I believe it is an invitation from Jesus that still stands.

Back to those three key things—the teachings of Jesus, his miracles, and (in a couple of chapters' time) what happened after his death.

The teachings of Jesus are still amazing, revolutionary even. We, of course, look at them with 2000 years of hindsight. To people hearing them for the first time, they must have been quite shocking. 'Love your enemies'[6]; 'Blessed are the meek, for they will inherit the earth'[7]; to mention a couple. He talked about God as an approachable father. He drove a horse and chariot through the crippling religious regulations of the day. He sought out the marginalized in society—women, the sick, and those with disabilities—and gave them status.

So, he was one of the good guys. Why not leave it there? After all doing good is no bad thing!

Unfortunately, Jesus didn't really leave us this option because his teaching was all bound up with some ideas that were not going to go down well. The old fashioned term is 'blasphemy'— equating yourself with God, claiming to be God—which is what Jesus did. There is more than one reference to Jesus upsetting people with his 'blasphemy'. In fact, it plays a very significant part in his trial and the events leading to his crucifixion.

You may have read a host of websites and articles that tell you that Jesus never said he was God or the Son of God. This, it is said, is just Christians reading things back into the stories that they believe. This shows fairly evident misunderstanding of what Jesus does clearly claim.

---

6   Matthew 5:44 (NIV).

7   Matthew 5:5 (NIV).

Jesus intentionally uses the Bible's name for God—'I am'. He deliberately forgives as only God can forgive—completely. He says that God is his Father and he and his Father are 'one', which leaves little to the imagination. When specifically challenged to say if he is the Son of God, he says, 'you have said so.'[78] Which is more like 'you got it guys' than 'Yes', but it amounts to the same thing. And then he takes on the challengers, 'Why then do you accuse me of blasphemy because I said, "I am God's Son"?'[9]

So, yes, Jesus did claim to be God.

Now let's look at the hands-on proof. I am not going to spend long on Jesus' miracles. There is a lengthy list of them. All four eyewitness stories of his life record them. Without exception they were not used to glorify himself. Jesus was not some ancient self-serving celebrity. They were used to help others—whether it be healing someone or as a way of teaching his followers. They show a power over nature and sickness and death that reflects God's power.

A bigger question is whether miracles can and do happen. In our scientific age we like explanations for everything. Anything that does not have a 'natural' cause makes us feel uncomfortable. Is it simply that Jesus did things that his followers interpreted as miracles because they didn't understand how the universe works?

It makes abundant sense that if, as Jesus claimed, he is God, then doing miracles would be something God can do. If God in some way, through his Spirit, nurtures the universe, then nudging it in certain 'natural' directions shouldn't be difficult. Is therefore a miracle something that happens anyway but at

---

8   Matthew 26:64 (NIV).

9   John 10:36 (NIV).

exactly the right time? Storms do subside after all. When Jesus calms a storm on a lake, it is the timing that is significant, not that the wind drops and the water stops going wild.

That said, Jesus' miracles cannot all be attributed to natural causes. Turning water into wine is just not going to be on the spectrum of scientific theories any time soon. So, maybe we are left with the uncomfortable truth that if Jesus did these things, maybe he was God after all.

Which brings us to the most uncomfortable part of the Jesus story.

8

Death and Hope

THE DEATH OF Jesus is one of the most influential moments in human history. It has been the subject of galleries of paintings, halls of musical compositions, libraries of books. It has been discussed and deliberated on from the day it happened. A Friday, as is happens.

Again the basic facts are fairly easy to jot down. Jesus meets with a few of his close followers to celebrate a Jewish festival in Jerusalem (the Passover, still celebrated by Jews today). Afterwards he goes outside the city walls, prays, and then he is arrested by the Jewish ruling elite. His friends abandon him. He is subjected to six trials which run throughout the night and early in the morning the death sentence is agreed—he is to be crucified.

An aside: Actually, the story is a bit more complex than this—so feel free to read it for yourself. (See the footnotes.[10]) What does become clear is that these illegal trials are not studies in best practice. There are a lot of vested interests being worked out. Jesus is innocent of the charges brought before him. All except 'blasphemy'. Which, of course, he is also innocent of if he is indeed God . . .

The Romans, as the ultimate ruling authority, take Jesus outside the city the same day and nail him to a cross. He hangs there for six hours and when the guards come to hurry up the dying process, they find he is already dead. To make sure they thrust a spear in his side and pierce his heart. The rebel rabbi has been dispatched for eternity. You would think the Jewish authorities would all relax and enjoy their day off—the *Shabbat*, Saturday. Oddly, they don't do this at all, but I will come back to that.

Another aside: Again, the story is far more detailed than this—read on in Matthew, Mark, Luke and John.

This you may already know. All these events take place in real time and real locations. We have enough archaeological evidence to place the main players in the story (like Pilate, the Roman governor, and Caiaphas, the Jewish leader). The settings are sites you can visit today. True we don't know exactly down to a GPS inch, but there are substantial clues to get us within a few feet, which for 2000 years ago isn't something to be sniffy

---

10   Matthew chapters 26, 27: Mark 14, 15; Luke 22, 23; John 18, 19.

about. And there are small details in the story which tally with practices and customs of the times.

Thousands upon thousands of sermons have been given on the aspects of what happened that day. And I am not going to entertain you with one of those. But it does raise the question of what I should highlight.

I have decided to go with one of the seven significant things Jesus said while on the cross. Partly because it is an interaction of Jesus with others. Partly because of what his words tell us.

Jesus was not the only person up for crucifixion that day. Two other men, criminals by their own admission, were being put to death. One is as bad as the crowd who mock a dying man. 'Aren't you our messiah? Save yourself. And save us while you are at it!'[11] The other seems to be a more honest lawbreaker and must have known something about Jesus already. He rebukes the first. The he asks Jesus:

Jesus, remember me when you come into your kingdom.

And Jesus replies,

Truly I tell you, today you will be with me in paradise.[12]

Here the story takes us deeper into gritty reality. Our quest to know who Jesus is moves away from a nice intellectual debate laid out in cosy words and neat arguments. We are drawn into a far more dramatic and dreadful setting for our discussion.

This cross conversation was no nice chat between a couple of blokes on a Friday afternoon. Jesus and this man were gasping

---

11   Luke 23:39 paraphrase.

12   Luke 23:39-43 (NIV).

for breath. Literally dying to find the strength to inflate their lungs one more time. Jesus had been whipped to within a few lashes of his life. Nails had punctured his wrists. The crowd below him were aggressive. Yelling insults. Mocking him. No doubt the other man had not exactly been treated with kindness either.

Yet a conversation does take place. One that those standing close by recorded. Were these two men in their pain-filled agony simply wasting their last breaths?

The criminal says some remarkable things in these few short words. Why should he expect Jesus to remember him? Not something to say to someone who is about to die, surely. And he talks about Jesus' 'kingdom'. What does he mean by that?

Jesus doesn't tell the man to stop wasting his last few moments on 'nonsense'. He talks about truth and paradise and promises something beyond expectation for a crucified criminal.

Paradise? Perhaps an old-fashioned word. 'Today you will be with me in heaven.' What are we to make of that?

Does anyone really believe in heaven? Doesn't life just shudder to a halt somewhere—hopefully later, rather than sooner? And that's it. A dark end to our story. Surely, this is just Christians (and others) hoping for the best when they die?

Well, let's be honest. Proving heaven exists is as difficult as proving God doesn't. You can read stories by people who have 'died' and seen the place, but while they are interesting, you may be inclined to disregard them. Trusting the experiences of others, as I have said, doesn't always come easy these days.

Let's head in another direction with this. By its nature, a spiritual realm where God lives is not going to show up on an electronic display, spotted by a passing satellite, or be found down some cosmic wormhole. Logically, if heaven is where God is and God is non-material, outside of time, then heaven is not going to be found in our familiar three dimensions or defined by our laws of physics and chemistry.

Perhaps it is difficult to get past this point. And you may need to hear me out without being convinced. Come along and see . . . believe it or not, there are some hopeful pointers.

The primary one is obvious—who is Jesus? If, as he claimed, he is God, then we have to take what he says seriously. More to the point is that it follows that this is a moral God speaking.

What right has Jesus to lie about the existence of heaven? What right has he to promise paradise to a dying man? If Jesus is no more than a man, then the kindest thing we could say was that he was trying to help someone out in a bad fix. If we wanted to be blunt, we would simply describe him as deluded or a charlatan, determined to keep up an immoral charade to the bitter end.

But Jesus' actions don't fit with those of a delusional character or a trickster. Logically, the best explanation is that Jesus is who he says he is. If he is God, he knows about heaven. If he is God, he has the right to offer places to people. And if he is God, then he has to be honest, truthful—because if he is not, then his own character is compromised.

So, what hope is Jesus offering? Life beyond death? If this was a stepping stone in a river, would I be willing to put my weight on it? Or would it keel over and pitch me headlong into the water?

Again, I am bound to say that dozens of years ago, I did just that. Like many millions of people, billions actually, have also done. I took that step of trust and discovered that the stone is rock solid.

Nevertheless, this is not simply, 'hey, follow me'. There is another extremely good reason for my faith.

But before that, let me finish my story—briefly . . .

Back in the day . . . 2

WHEN I LEFT off, I was at Wheaton College, poring over maps and praying about where in the world God wanted me to go. In the summer after high school I had been down to Mexico with a youth team from my church. There I had briefly met another young guy, George Verwer. George was to become the founder of an organization called Operation Mobilisation, which Yvonne and I would later join. Meanwhile, however, I began to give a lot of attention to Turkey.

In 1961 I decided to take a year out of college. Dale Rhoton, another Wheaton college student, had been encouraged by George to set up a Christian correspondence course in Istanbul. Dale was delayed, so George suggested I go ahead. The cheapest way to get to Europe with as much stuff as possible was the transatlantic liner, the Queen Elizabeth. So days before my twentieth birthday I turned up at the New York dockside with a ton of literature, second-hand clothing and supplies, and one small suitcase of my own possessions, ready to set sail for France.

In Europe George joined me with another idea. We would find a van, load it up with Christian literature and head for the border with the Soviet Union—at that time a Communist, atheistic regime, where the church was being persecuted. Things went smoothly getting our Opel station wagon into Czechoslovakia.

We were searched, but the literature wasn't touched. From there we made it into Ukraine.

All was going well until George tried to hand a damaged copy of John's story of Jesus out of the wagon window. Caught by the breeze the pages fluttered off down the road. Whoever picked it up was not thrilled about what they read. They handed it in to the authorities.

Heading to Kiev we got stopped near Rivne. Armed police were standing in the middle of the road, brandishing Kalashnikovs and accusing us of being spies. Ending up in a gulag labour camp suddenly looked a distinct possibility. We were questioned for two days before we were escorted to the Czech border—a motor cycle up front leading the way and one behind. We made the papers. Both the Soviet Pravda and American International Herald Tribune reported our detention and expulsion.

My real focus, however, was Turkey. As a fiercely Islamic country the Christian church was miniscule. The believers that did exist there were vastly under resourced. I set up home in Istanbul and Dale and I started producing Christian literature for distribution. Then I got together a small group of Greek and Armenian Christian kids and ran envelope addressing 'parties', preparing to advertise a free correspondence course. By autumn 1962 we had maybe 10,000 letters to mail, maybe more, and we put them in the Istanbul post all at once!

Guess what? I made the papers again. Everyone in Istanbul knew about the foreign guy who was behind the headline 'Christian propaganda hits city'. We had been wondering if we would get—been praying for—one response. We needn't have been anxious. There were scores of replies. However, as I went to open our PO box, I heard a voice behind me, 'Will you come with us . . .'

Once again I was hauled in for questioning. Another two days. Thankfully, this time I wasn't shown the border gates and, while the advert replies were confiscated, we had found out that there were plenty of people in the city who wanted to learn more about this prophet they found in their Qur'an—Isa—Jesus.

The next time I was in the local newspapers I was getting married! In December the following year Yvonne came to Turkey to join me and in January 1964 we were married in a Dutch chapel in downtown Istanbul. Despite the setback, I continued the mailing of Christian literature, while Yvonne taught English in the city. And two years later I had my third and final brush with authority.

I guess the powers that be worked out that getting rid of me was the quickest way to stop the correspondence course functioning. One afternoon I went down to the government security offices to renew Yvonne's visa and was arrested. A proper criminal— fingerprinted and photographed, front and side shots like you see in the movies. Then, handcuffed to a policeman, I was to be taken to the Greek border—pronto, no time to pack. I asked if I could at least inform Yvonne. So I found myself being marched up the lane to our flat in handcuffs—let me tell you that it is just embarrassing having the neighbourhood women watch you being escorted by a policeman all the way to your front door. Yvonne wasn't in. So, I left a note, made the policeman a cup of tea, and we headed for the border. They slapped a huge X on my Turkey visa and told me to walk over the bridge into Greece. From there a Greek Christian put me on a train to Frankfurt and six weeks later Yvonne was able to join me. When we tried to return to Istanbul, we found we were on an official 'Do not allow to enter Turkey' list. And that, sadly, was the end of our residency.

Since then Yvonne and I have maintained our contacts with the church in Turkey, visiting from time to time, still concerned to see them well supported in a difficult situation. So, over many years we have continued to provide literature. Alongside this, I have been involved in a number of film projects—feature films and documentaries—sharing what I have learned about God and about Jesus.

Why am I telling you all this? I want you to understand that deciding to follow Jesus at age 8 was not just an impressionable childhood whim. It was a decision that has affected my whole life. Seventy years later I am as passionate about telling others, you included, about what I believe and why, as I was when I raced around our neighbourhood in Nebraska to invite my friends to my Mom's 'Good News club'.

9

Getting up on Sunday

For years I have been getting up every Sunday morning and going to church.

An aside: In case you had not noticed, not all Christians do this. Some choose to meet on Saturday, in keeping with the Jewish day of rest. Others pick another day to fit with their work schedules.

There is good reason for this. It is a day to celebrate. Sunday was the third day after the death of Jesus. The day he said he would rise from the dead. In the realm of 'prove it', is there a better way to establish the existence of life after death than to demonstrate the truth of it?

This is another remarkable thing about Jesus. His teaching was revolutionary. He did miracles. He made a claim to be God. And he made it clear that he expected both to die and to come back to life again. Which is a dilemma all of its own.

Whatever picture we have of God—whether we believe in

God or not—I doubt that for most people God is capable of dying. That is the fate of animals and humans, not God.

Here we must engage with theology—the study of the nature of God. First off, it goes without saying that we may have views on what God is like, but that doesn't mean they are right. We can easily try to restrict God to what we think he should be like, what we think he should/shouldn't do. But actually we would be better listening to or seeing what he reveals. Many of us have probably felt that we have been misrepresented and thought, 'if only people knew me better . . .' Perhaps God feels the same way!

So, is it possible for God to die? Well, this is all bound up with who Jesus is. The Bible and Christians don't claim that Jesus is some kind of other-worldly God floating around in human form; a really realistic ghost, who cries, 'you can't hurt me', as a spear is pushed into his side. The Bible describes Jesus as fully God. And fully human. A tricky equation to solve, if you only work in 3D, of course.

The resurrection of Jesus doesn't just leave us discussing theology. The issue is whether he did *actually* rise from the dead. There is a need for some decent evidence which stands up to scrutiny if you are going to make claims like this. And I am not going to disappoint you. Once again I can point to history as hard evidence that you have to explain away if you want to reject the reasons for my faith. Perhaps it is best if I approach this from a non-believer's point of view.

An aside: I cannot condense all the evidence satisfactorily into a couple of paragraphs. There are numerous books on the matter, a few of which are mentioned at the end of this one, which provide a more comprehensive account.

The obvious thing to say is that the resurrection just isn't possible, so it cannot be believed. And the obvious response is that nobody is claiming that this happens 'naturally' or frequently enough to be an observed 'fact' of the type science likes to see, record and thereby validate. Declaring it impossible fails to engage with the story in all its facets.

OK. Next up and more pragmatic. Maybe people got the wrong tomb. They simply turned up a couple of days later and, hey, they couldn't quite remember which one was the right one. This one happened to be empty and history was altered.

This begs an awful lot of questions. Jesus was lain in the tomb belonging to a guy called Joseph, by Joseph himself. Hard to believe he couldn't remember which tomb he had recently prepared for his family and himself. Then there were women close at hand who wanted to put spices on the body. Could they all forget? And why didn't the authorities, who had posted guards at the tomb, rock up with crushing news that Jesus was still dead and buried a couple of graves along?

Actually, here is a good time to talk about the local authorities. One of the reasons they didn't rest too well on the Saturday after Jesus' death is that they knew that Jesus had promised to rise from the dead. So they insisted on guards being posted at the tomb. They also had a seal placed across the entrance. Nobody was stealing this body.

The next comeback is that maybe people were mistaken. After all the followers of Jesus must have wished he was alive—grieving people do. Here we are into mass hallucination on a grand scale. The followers of Jesus claimed to have seen Jesus on more than one occasion, for lengthy periods of time, not fleeting glimpses as a ghostly figure flickered across their path or slipped inconveniently around a corner. They said he talked with them, ate with them. Some of these people had spent three years with him, hardly likely to mistake someone else for him.

The stories even tell us that it was in his mannerisms that he was recognized. And how many followers had this experience? Not just a few—five hundred.

OK, one more go. (And as I have noted others have gone into the evidence far more thoroughly than this.) Perhaps all these folk were gullible, impressionable types; easily conned into believing Jesus really was alive. Couldn't this have been a skilfully executed scam?

It is difficult to see exactly what such a scam would have achieved. And who was running the scam? Jesus? Well, he had been beaten nearly to death, nailed to a cross and had a spear pushed into his heart. Tricky to organize that hoax with the hostile Jewish authorities in attendance and hard to see what the Romans were getting out of this.

What about the main group of followers? They had a lot to gain. Instant fame. Book contracts. Their own TV shows. And all the perks of stardom as the leaders of this new religion they didn't have a name for. If so, they got it badly wrong. Nobody made a fortune. One of them, James, was soon put to death by a local ruler for proclaiming Jesus was alive. And when it was obvious that plan A wasn't working and the others were facing imprisonment and the death penalty themselves, why, oh why, didn't they just plead guilty to deception.

Which brings me to the problem of the 'deceived'. We may tend to think that everyone in the past was somehow incredibly naïve. That they were so superstitious that anyone could turn up a silly story and get it believed. Only in our sophisticated, scientific era has sense prevailed.

Well, the reported resurrection of Jesus was certainly in the realms of silly stories if you wanted to generate a following. Pushing the boundaries of anyone's beliefs you would have thought.

But let's leave that there, because we know that not everyone just jumped at the chance to believe it. Not surprisingly, when you think about it, the responses to the news varied; an assortment of reactions—not unlike our generation after all!

Some did simply trust others. Their friends said they had seen Jesus, why should they not believe them? Other followers went to the tomb to see for themselves. One, we know, finding Jesus' grave clothes but no body, needed to look no further. One of Jesus' closest followers, Thomas, was not impressed with the reports. He wasn't going to believe it until he had pressed his finger into the nail wounds and felt the scar of the spear cut in Jesus' side. I guess, he was both shocked and delighted when Jesus turned up and invited him to do just that.

And so it goes on. Evidence, like this, is presented. There are those who set their barriers out good and proper, come what may, those who believe readily, and those who need time to decide.

Of course, there are plenty of people who choose not to engage at all. Better to keep their heads down, make the best of this life, right? They will worry about life after death if and when the time comes. Except, as we know, death isn't an exact science.

Then, there are those who would like to believe it, but the pressure of others holds them back. Which is a great shame, because this is serious stuff—eternally serious, I believe. But it happens.

But in the main I hope people want to think about it, work it out for themselves. So they read, go along to a church, watch a few videos, listen to both sides of the debate. Nothing wrong in that.

10

This Book: The most valuable thing that this world affords

THERE IS A very significant phrase spoken at the crowning of the head of state in Britain, the coronation of a new queen or king. The monarch is given a copy of the Bible and informed that what they have in their hands is 'the most valuable thing that this world affords'. This is not a recent innovation. The words go back to April 1689. Prior to that monarchs would kiss the book after they had made an oath of commitment with their hand on the Bible—something American presidents have also done.

Whether this is now simply tradition or carries more weight than that, whether the individuals concerned owe anything to its contents, it has to be said that somewhere along the line the Bible has achieved a significance given to no other book.

It has also inspired men and women to amazing acts of bravery and self-sacrifice. It has helped shape the beginnings of modern science. It has been a resource to be plundered for the most magnificent of paintings, music and literature. It has been carried into battle. It has laid the foundations for legal systems and the rise of democracy. It is the bestselling book of all time. It has been a guide and comfort to Christians across every nation and to many who wouldn't call themselves Christians. Sadly, it has also been at the centre of bitter controversy and in corridors of power used by people to inflict great misery on others. It is not a book that can be ignored!

Perhaps it is best if I first describe the features of this book, because unsurprisingly it is no ordinary book at all.

The basics are easily dealt with. It is a book of books; 66 of them, in fact. It is divided into two major segments; the Old and the New Testaments. The Old contains 39 books, the New 27. The Old is the Jewish scriptures written prior to the time of Jesus, so these are the words he would have been familiar with and used in his teaching. The New came into existence after the death and resurrection of Jesus, the result of the writings of the first Christians.

So far, so good. But then it gets a little more complex. Written over hundreds of years before Jesus lived and a few decades afterwards, the Bible is a collection of ancient writings bound together into one large volume. We have names for nearly all the New Testament writers, but only some of those who wrote books of the Old Testament.

It is a mix of languages: Hebrew (Old) and Greek (New), with a few phrases of Aramaic thrown in for good measure.

It is also a library of literature styles; history and poetry, legalities and letters, philosophy, prophecy and proverbs. Different books adopt different genres or combine more than one.

And, strange as it may now seem, it didn't slide off a scribe's writing desk as a complete compendium until around 400 years after Jesus was born—technology just wasn't ready for such a large book to exist. (And just in case you were wondering, it took another 1000 years or so to add in the chapter and verse numbers that we use today.) In the beginning it was mainly a collection of scrolls (Old) and small papyrus books called codices (New). And before that sections of the Old Testament were passed on orally from one generation to another before anyone got around to writing it down.

Does this all seem a bit haphazard for such an influential book?

The easy response is that it isn't how the Bible got to be the book it is that ultimately matters. It is what is written between the covers that makes the difference. And this is true. But hardly fair! How can I ask others to put faith in the writings of a book without some justification for its creation and its trustworthiness?

11

On Word

THE HISTORY OF the Bible—how it came into existence—is like other topics covered in this book, far too detailed for the space here. What I want is mainly to avoid the dry history and get to the issue of authenticity. And also I want to major on the New Testament, because this is the part that tells the stories of Jesus and expands on what the first Christians tell us about him. However, that said, it is important to realize that for Christians the Old Testament is also integral to understanding who Jesus is.

There are dozens of prophecies in the Old Testament that Jesus fulfilled as the expected 'Messiah'. The New Testament writers quote from the Old Testament a lot, using its verses to present Jesus to their readers. Jesus' teaching, as I have said, came out of the Old Testament writings, even if he often gave a radical new understanding of what it said.

So, Old Testament first . . . but briefly:

A large chunk of the Old Testament tells the history of the Hebrew people—the Jews, but the story starts long before these 'people of Israel' emerged as an identifiable group.

There is an interlude of poetry and proverbs, followed by a long final section of the words of Israel's prophets. By Jesus' day these Hebrew writings were available on scrolls kept in synagogues, each copy written out by hand.

Going back to the early origins of some of the books, we tend to be a bit sniffy about oral traditions. It is difficult for us to imagine a world where every last iota of information is not recorded in 'hard copy'. We play the Telephone game and laugh at the strange message that emerges at the end of the line. But then we don't pin our existence on an oral system. If our fragile society depended entirely on getting the message from one end of the line to the other accurately, wouldn't we make sure the message was passed on properly? Small communities well versed in their stories were jealous guardians of their traditions. This *was* their 'hard copy'.

Talking of telephones, it isn't that long ago that people remembered telephone numbers and if you needed to phone someone, you could ask someone for the number and they would reel it off from memory. I guess most young people would find that hard to believe as they scroll down their mobiles and read it out.

In time the old oral stories were written down onto scrolls. Then, newly written books were added to the collection. Together these became the sacred writings of the Jews, and then included in the Christian Bible.

If we are superior about the reliability of oral traditions, we also like to point to the permanence of print. Once a book has rolled off the press the words on the page are static, difficult to change. The age-long process of human scribes, generation after generation, copying from one manuscript to another was bound to introduce errors into the text—intended or not.

There is no doubting that ancient manuscripts do contain errors. Nobody would pretend that we have a Bible that has every word perfectly preserved in stone from the moment it was spoken or first written down. (This leads to a question about what God is doing in all this. I'll come back to that.)

We need to give some real credit to the old scribes though, rather than dismiss them as a bunch of grade ten teenagers copying out lines from a textbook as a punishment for some misdemeanour. The whole point of transferring the text from one document to a new one was to do so accurately. Particularly when it came to sacred texts! And you can bet that someone was checking.

Moving to the New Testament. The obvious advantage is that the books were all written in a relatively short period. Compared with the Old Testament things had moved on a lot in terms of communication. And we have a lot of evidence for the New Testament's authenticity.

The books of the New Testament can be divided into four categories:

- Stories of Jesus (*Matthew, Mark, Luke and John*).
- A history of the early church (*Acts*).
- Letters to churches in (today's) Turkey, Greece and Rome.
- The book of *Revelation*—an interesting book to finish with!

All of these writings are first century AD. Jesus died around 33 AD, so this places them in a 60-70 year window after his death. All but one book has the name of the writer attached to it, either in the text or ascribed.

Taking each category in turn:

- Matthew and John were close friends of Jesus. Luke and Mark were among the first Christians, relying on eyewitness reports.

- *Acts* was also written by Luke, in part a description of his own experiences.

- The letters were written mainly by Paul, who knew the close friends of Jesus, with other letters written by some of Jesus' closest followers.

- *Revelation* was written by John.

What is essential to understand is that the stories of Jesus are records of eyewitness accounts. There was no telling of tales passed down to from generation to generation. Names of individuals are given. Incidents are coloured with information. Details reflect what we know of the time. Reliable history and verifiable geography are integral to the stories.

One of the practical reasons why Christianity spread quickly was that its followers took advantage of a newish technology available to them. Interestingly, the first Christians abandoned the Jewish reliance on scrolls. Small papyrus books—codices—were cheaper and quicker to produce, and easier to read.

Allied to this, Roman road and sea routes opened up great highways for communication. Sure it took a few days to get a message from one side of the empire to the other, but it was only a matter of days. As a consequence stories and details could be checked more easily.

Of course, you may have discovered that plenty of people with their own agendas argue that these initial writings were distorted; their transmission relying on the accuracy of handwritten copies. (The printing press arrived in the 1400s.) Or even that Christianity was really a product of later generations of Christians rewriting the script to suit their own purposes.

The proof of this pudding seems to me to be in the ancient manuscripts we have of the New Testament. My basic question is this: is our Bible the same as the manuscripts we have from the second century onwards? If so—as is the reality—where is the evidence of distortion or Christianity being a product of another generation other than those who knew Jesus or were his first followers?

An aside: I was the producer of a 30 minute documentary, The Jesus Accounts: Fact or Fiction?, which deals in detail with this question. You can find it online on Youtube, along with other productions I have been involved with.

Two more comments on this:

Did you know that even if we had no ancient manuscript copies of the New Testament, we can still replicate a large proportion of it from quotations used by church leaders in their writings from the second century onwards?

Did you know that we have a considerable number of early New Testament manuscripts, thousands in fact? Far more than any other ancient writings, other than those of Homer who comes a distant second in the list. And, while new discoveries will continue to be made, the time gap between the ancient New Testament copies we have and the date of the originals is considerably shorter than for other ancient writings where the gap can be hundreds of years.

People are quick to point to the differences between these ancient manuscripts—variants—as evidence that our New

Testament is in error. Therefore, they argue, it cannot be trusted. The point is poorly made. Christians are not arguing that the variants are not there, nor that they can be ignored. Finding a variant is a sure way to have New Testament scholars scuttling back to their desks to work out the original text!

Much is made of the number of variants in the ancient texts and again no-one is denying this. However, these 'errors' are more often than not simply trivial—for example, spelling mistakes where the meaning is 'obvous', if you see what I mean? Certainly none of these variants alters the basic message of the New Testament, that Jesus is the Son of God, who lived, died and rose again, because God loves me and he loves you.

The fact that we have so many manuscripts is good news. This allows us to have confidence in a Bible that is based on the reliable, not the random. As a comparable example, imagine a room full of people needing to know what the time is as they wait for a train. If there is only one watch between them, how do they know whether it is correct? The more clocks on show, the better!

12

God's Word

THAT LAST CHAPTER was a little lengthy, so I'll try and keep this one short. The topic, however, is also very important to my faith.

I would say that the Bible is God's word, one of the key ways in which God communicates with us. This, of course, puts it in a different category than simply a fine piece of literature with an interesting history and ongoing significance.

In its own words:

All Scripture is God-breathed and is useful for teaching, rebuking, correcting and training in righteousness, so that the servant of God may be thoroughly equipped for every good work.[13]

It is those words 'God-breathed' that give *all* of the Bible a category of its own. So what does that mean?

An alternative translation of the original Greek words is

---

13   2 Timothy 3:16-17 (NIV).

'inspired'. The Bible was not dictated by God word for word to writers who sat like robots with their reed pens in hand. The words did not appear as if by magic as they scribbled across the papyrus or parchment pages.

God was perhaps more like a driving instructor or sports coach; closely involved, providing direction, leaving his imprint on the result of their endeavours. Being inspired takes nothing away from the creativity and craft of the writer. But it does acknowledge that there are deeper purposes at work.

It doesn't take much research to discover that there is a view that the Old Testament is the story of a fierce, judgmental God and the New Testament in Jesus presents a kinder face. At first glance this is not unreasonable. The Old Testament histories, prophesies and poems deal with a lot of conflict, within families and across borders. Tragic and awful things happen, apparently sanctioned by God. Jesus, on the other hand, doesn't go to war. He speaks peace and sacrifices himself. Yet, for Christians the God of the Old and New Testaments is the same God. There is no division. Deeper study uncovers a kind, loving God of mercy in the Old Testament, providing a counterbalance to the popular view.

However, I think really this goes back to my first point. People wrote the pages of the Bible. They did so within their own times and their own understanding of events and outcomes. Interestingly, many of the Old Testament Bible stories which receive a hard press today were readily accepted, even rejoiced over, when I was young. The world then, at war with itself in the 1940s and through the subsequent 'Cold War', saw things differently!

The Bible is not a simplistic read. It invites us to study, to grapple with its stories and its deeper meaning. God, I believe, has given us intelligent minds to do this. It cannot be reduced to a rigid mantra to be repeated endlessly by every culture and through every millennia. The Bible is God's word to each generation, regardless of ethnicity, location and status; each one delving into it and discovering truths. In short, it is God-breathed and therefore has life.

*   *   *

Psalm 119—the longest psalm of all, near the centre of the Bible—has a lot to say about God and his word and it is a good place to finish this chapter.

> *Your word is a lamp for my feet,*
> *a light on my path.*

> *The unfolding of your words gives light;*
> *it gives understanding to the simple.*[14]

---

14   Psalm 119:105, 130 (NIV).

13

God Calling

SOME WHILE BACK I made a comment about God:

What kind of God do I believe in?

A good God, who communicates with the world. Now the best bit: history, I believe, shows this supremely in the life, death and resurrection of Jesus.

Having made a case for the existence and character of God from science, and used history to look at the life, death and resurrection of Jesus and where the Bible fits into that, I hope you can see that there is a connection to you and your life.

The stories of Jesus detail his remarkable declaration to be God, borne out by his extraordinary actions; doing miracles, forgiving wrongdoing. His death is no ordinary first century Roman crucifixion; even in the midst of great distress he takes a gasping breath to promise eternal life after death to a dying criminal. And how can he do this? Because, as the Bible says:

> *[He] came not to be served but to serve others and to give his life as a ransom for many.*[15]

His resurrection sets him apart in world history. And out of this life, death and resurrection comes a movement—Christianity—that has a profound effect on the lives of billions of individuals, me included.

What I want to emphasize is that Jesus is not just an aberration in time and space. As though God got bored of nurturing the universe and decided to have a go at being human. Or felt it would be good to get to know a few folk. The landlord who moved in for a while, found the place wasn't as nice as he expected and so moved out again.

I believe that there is so much more to this. I am going to stick with the word 'hope', although you would maybe find other Christians quoting another couple of Bible verses at this point:

> *We have left God's paths to follow our own. Yet the Lord laid on him the sins of us all.*[16]

> *God so loved the world that he gave his one and only Son, that whoever believes in him shall not perish but have eternal life.*[17]

---

15   Matthew 20:28 (NLT).

16   Isaiah 53:6 (NLT).

17   John 3:16 (NIV).

Which is a good way to go, but right at the beginning I mentioned I need to give reason for the hope I have. So, 'hope' it is.

> *An aside: One of the terrific things about the Bible is that it contains some amazing prophecies about Jesus and what he would do. The first of these verses—'We have left God's paths . . .' is part of a much longer passage in the Old Testament, the words of a prophet named Isaiah. (The whole of Isaiah chapter 53 is well worth reading if you can get hold of a Bible.) Isaiah lived around 700 years before Jesus was born.*

The hope I am talking about isn't 'I hope it won't rain'—a kind of wishful thinking, lacking certainty. My kind of hope is more like a well-constructed tower securely built on rock.

Where do I get this kind of hope from?

- From a good God, who I can trust because he is God and he is good.
- From a God who can and does communicate.
- From a God, who has not remained remote, speaking faintly, but who speaks clearly.
- From a God who has engaged fully in my world, living and dying alongside humanity and then rising to life again to bring the good news that hope exists in this universe.

For me, Jesus, who he is and what he has done, remains the ultimate expression of God at work in the world. That is why I follow his example, seek to live my life as he would, continue to explain to people what I believe and the good reasons I have for my beliefs.

Way back in the first chapter I also said that going back to the basics of what we believe is a great opportunity to see where we have come from to get where we are, and to decide where we are going next. So maybe that is where you are—deciding where to go next. Maybe just to get another coffee. Or maybe you would like to head in a different direction than the kitchen?

Can I suggest you look at the books I have listed below or watch the documentaries I have been involved in making. They go into much greater depths than I have been able to.

Maybe you want to get hold of a Bible and read what it says for yourself. Bibles are easy enough to pick up free online. Or if you prefer a hard copy, any good bookstore will have a one. I would go for a modern translation—the New Living Translation (NLT) or the New International Version (NIV). And start in the New Testament with one of the stories of Jesus—*Matthew, Mark, Luke,* or *John.*

Getting along to a church and meeting other Christians would be another way forward. Tell them what you have been thinking about and listen to their stories of God at work in their lives and how and why they have committed their lives to following Jesus.

And from there to where?

Finishing with Something

WE HAVE COME *a long way. From the far reaches of space and time to a logical conclusion that God exists. From there to understanding something about a God who is good. And with that a knowledge that God cares about us. That he communicates with us. And that he communicates particularly through the life, death and resurrection of his Son, Jesus.*

So, if you really want to understand why I am a Christian, it's all very straightforward—you need to understand who Jesus is and what he has done for you.

God bless,

Roger

Things to read

The Historical Reliability of the Gospels, Craig L. Blomberg, Apollos, 2007

On Guard, William L. Craig, David C Cook Publishing, 2010

Reasonable Faith, William L. Craig, Crossway Books, 2008

I Don't Have Enough Faith to be an Atheist, Norman L. Geisler & Frank Turek, Crossway Books, 2004

The Reason for God: Belief in an Age of Scepticism, Tim Keller, Hodder & Stoughton. 2009

God and Stephen Hawking: Whose Design is it Anyway?, John C. Lennox, Lion Hudson, 2011

God's Undertaker: Has Science Buried God?, John C. Lennox, Lion Hudson, 2009

Gunning for God, John C. Lennox, Lion Hudson, 2011

The Resurrection Factor, Josh McDowell, Authentic Media, 2005

Just Six Numbers, Martin Rees, Weidenfeld & Nicholson, 1999

Can We Trust the Gospels, Mark D. Roberts, Crossway Books, 2007

The Case for a Creator, Lee Strobel, Zondervan, 2014

The Case for Christ, Lee Strobel, Zondervan, 1998

The Case for Faith, Lee Strobel, Zondervan, 2000

Jesus Among Other Gods: The Absolute Claims of the Christian Message, Ravi Zacharias, Thomas Nelson, 2010

Jesus Among Secular Gods: The Countercultural Claims of Christ, Ravi Zacharias & Vince Vitale, Faithwords, 2018

Seven Days that Divide the World, J. C. Lennox, Zondervan, 2011

Go online:

Youtube—Goosewing Productions:

https://www.youtube.com/channel/UCqrutnY7Ymte Etjdc9HcEsw

The Jesus Accounts: a 30 minute documentary on the reliability of the stories of Jesus (Producer: Roger Malstead; Director: Ronald Clements)

* * *

Bible references:

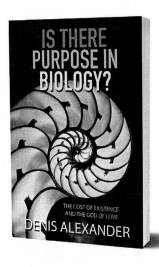

Safely Home by Randy Alcorn

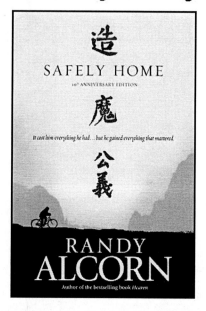

Randy Alcorn wrote a novel about the persecuted church in China that changed the face of Christian fiction. Alcorn wove a tale of intrigue and hope as he took readers on a journey from corporate America to the persecuted church in China. Today, the novel challenges readers to consider if they are prepared to suffer and die for their faith, while reminding them of the hope of Heaven and the importance of living with eternal matters in their daily lives.

Learn more at www.epm.org/safelyhomebook

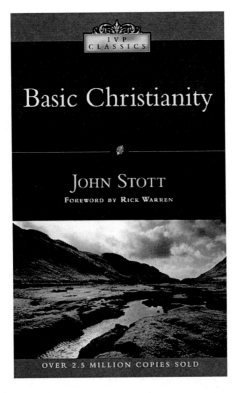

Basic Christianity

JOHN STOTT

FOREWORD BY RICK WARREN

OVER 2.5 MILLION COPIES SOLD

Named a "Book of the Century" by *Christianity Today* "If Jesus was not God in human flesh, Christianity is exploded," writes John Stott. "We are left with just another religion with some beautiful ideas and noble ethics; its unique distinction has gone." Who is Jesus Christ? If he is not who he said he was, and if he did not do what he said he had come to do, the whole superstructure of Christianity crumbles in ruin. Is it plausible that Jesus was truly divine? And what would that mean for us? John Stott's clear, classic book, now updated, examines the historical facts on which Christianity stands. Here is a sound, sensible guide for all who seek an intellectually satisfying explanation of the Christian faith. www.ivpress.com

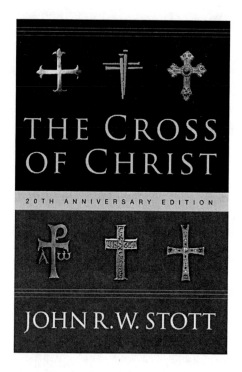

THE CROSS OF CHRIST

20TH ANNIVERSARY EDITION

JOHN R.W. STOTT

The work of a lifetime, from one of the world's most influential thinkers, about the heart of the Christian faith.

"I could never myself believe in God, if it were not for the cross. . . . In the real world of pain, how could one worship a God who was immune to it?" With compelling honesty John Stott confronts this generation with the centrality of the cross in God's redemption of the world—a world now haunted by the memories of Auschwitz, the pain of oppression and the specter of nuclear war.

Can we see triumph in tragedy, victory in shame? Why should an object of Roman distaste and Jewish disgust be the emblem of our worship and the axiom of our faith? And what does it mean for us today? www.ivpress.com